COLORADO MINERAL COLLECTING

1957 TO 1972

Colorado Mineral Collecting
1957-1972

By Donley S. Collins

Published
Colorado Springs, CO
2025

Colorado Mineral Collecting: 1957 To 1972

Published in Colorado Springs, Colorado by Donley S. Collins

ISBN 979-8-218-82949-0

Contributing editor: Robert W. King

Cover Design and interior formatting by:
Elizabeth Harper at EHR Ltd.

elizabethharperreads@gmail.com

Dedicated
to
E.B. Eckel

INTRODUCTION

In 1956, I accompanied my aunt, Audry Patton, to a Colorado Springs Mineralogical Society (CSMS) meeting that was held once a month at the Alexander Film Company located on north Nevada Avenue in Colorado Springs, Colorado. I was about 8 years old at the time, and this was my first attendance to many meetings and field trips to follow. During the summers from 1957 through 1970, CSMS conducted field trips to popular mineral localities throughout the state. These monthly trips were a family event that brought many club members and their friends together. It was also a time for picnics as well as a chance to visit out of the way scenic locales. Members would meet at a prearranged starting point and provide people without vehicles with a ride. These trips, usually lasting a day, offered me, as well as amateur mineralogists (rock collectors), a chance to understand mineral and rock associations as well as their occurrences. At that time, it was a treasure hunt that could satisfy any amateur or professional collector.

Since 1972, many of the CSMS's classic and general public mineral locales have been lost; a result of either over hunting, land development, claims, or by local and federal government regulations. At collecting sites still available to the public, the once abundant near surface mineral deposits have become difficult to find and the quality and size of specimens are lacking as compared to minerals deposits found during the late 1950's through 1972. As a result, in some cases, heavy machinery is now used to uncover subsurface deposits.

Sadly, many CSMS and non-CSMS collectors that I have known during the late 1950s through 1972 have passed and many of their spectacular finds have been lost: a result of lack of labeling, missing labels, improper storage or worse, sold, thrown away or donated to museums by disinterested relatives. For instance, through the years, I have found several collections being sold at garage sales; the sellers not knowing what they had. At other times, I have witnessed entire collections, having wonderful minerals, just thrown into waste bins by disinterested relatives. On the other hand, some collectors donate or will their collections to museums thinking that their entire collection will be preserved, but museums rarely retain an entire collection because of limited space. After selecting portions of a donated collection, museums often sell, trade or donate the remaining minerals of that collection to other institutions or to private individuals.

This text documents minerals from the popular collecting areas in Colorado from the late 1950's through 1972. All of the minerals shown in this book were collected in this time period. Amateur guidebooks referenced and used during this time, included "Colorado Gem Trails and Mineral Guide", Richard M. Pearl, 1958 (third printing) and the less known "Minerals of Colorado--100 Year Record ", (E.B. Eckel, 1961) which was sold for one dollar in 1961 and through the 1980's by the U.S. Geological Survey. In addition, I had the good fortune to have known advanced collectors of CSMS with much experience that included: Clarence Coil, Raymond (Ray) Ziegler (Photo 1), Stuart (Stu) E. Hays (Photo 1), Gale Zinn, Max Fillmore, and Chris Christenson, all of whom had the patience to teach me their collecting methods.

Photograph 1. My friends (from left to right)
Raymond Ziegler, Stuart Hays, and George Fisher
(past members of CSMS).

COLORADO SPRINGS AREA

Within the Colorado Springs area (El Paso County), an alabaster (granular variety of gypsum) bed, a sandstone bed with micro-double terminated quartz crystals, and a dark blue celestite occurred in and adjacent to the Garden of the Gods Park.

The whitish-pink alabaster formed a massive vertical bed on the east side and juxtaposed to the Red Lyons Formation within the Garden of the Gods Park. Once carved into light houses with added lighting, these carvings as well as other carved topics were abundant in curio and tourist shops in Manitou and Colorado Springs.

East of the alabaster bed was a vertical, white, very friable, fine-grained quartz sandstone bed (part of the White Lyons Formation?). The sand grains forming this bed are white, clear, double terminated quartz crystals, a result of diagenesis. These grains were perfect for "micro-mount" collectors. Unfortunately, sometime after the 1970's, remodeling of walkways through the park has eliminated both the alabaster and white sand outcrops.

In late 1968, I managed to save $350 for a mineralogy class taught by Richard M. Pearl at Colorado College. At that time Professor Pearl was updating one of his mineral guidebooks and later he offered an "A" to anyone that could find a celestite specimen from a locale just north of the Garden of the Gods Park. He then produced for the class a beautiful example within a reddish-brown nodule measuring about 5-inches wide and 6-inches long, containing dark blue, less than 0.5-inch long, gemmy celestite crystals. It was on a day near the end of the semester--mid-December--when he made this offer. That evening, Colorado Springs received over a foot plus of snow which did not melt until well after the semester ended; thus, preventing anyone from searching for that site and claiming the reward. A few years later, I described this site to George W. Fisher, a late member of the CSMS. George (an avid collector) searched that general area and found an example of this mineral, but nothing like that found by Richard M. Pearl. Since Fisher's find, it is believed that this celestite locale is depleted. I have searched the Colorado College mineral exhibits and did not locate Dr. Pearl's celestite sample. It is said that Dr. Pearl's collection was sold, but I could not confirm this. This is an example of a depleted occurrence resulting in a lost or forgotten locale.

In the late 1950's collecting was, and is currently, prohibited within the Garden of the Gods Park as well as in the adjacent areas, a result of private and new land development. If anyone attempts to find this locale, be sure to check if permission is required to hunt that area.

One of the most outstanding selenite locales in Colorado is now buried under a bridge support for the Highway 24 overpass located near the intersection of Circle Drive and Fountain Boulevard in Colorado Springs.

In 1962, an eighth-grade classmate showed me an amber colored, high-lustered selenite "rose" which he had found in the bank of Spring Creek, just to the east of the Evergreen Cemetery. He had no idea what it was, but he knew that it was "neat." The rose was palm size with "petals" about 0.5-inch long radiating from the center of the rose structure. Next day after school, he and I bicycled to the site. What I found were abundant selenite crystals in a very wet adobe clay exposed in the south bank of an east-west trending creek. For several weeks I rode my bike with my fossil pick in my belt and carried as many crystals home as would fit in my WW2 backpack. Soon after, my eighth-grade math teacher, Robert (Bob) King (Photo 2), asked why I had not been coming in after school to play chess. When I showed him why--he soon joined me. This was Bob's first collecting trip with me. This started a long friendship and collecting partnership that has continued off and on to the present.

Photograph 2. Donley S. Collins (left to right),
Robert W. King, and Andy Kowalski in 2021.

While collecting the selenite crystals, Bob and I found that the crystals were of two distinct colors: grayish-white (white) and amber (Photo 3) with each having slightly different structure variations as represented by their "rose" and single monoclinic crystal form. The white "rose" variety (Photo 4) had a more prominent "rose petal" development and a poorly developed single monoclinic crystal form. The white roses ranged from less than 0.5-inch to over 3-inches in diameter, whereas the less common single monoclinic crystals ranged from 1- inch to well over 3-inches long and about 0.5-inch thick. The white variety crystals seemed to occur in a zone a little east of the amber colored selenite occurrence.

Photograph 3. Color contrast between the two varieties of selenite crystals found along Spring Creek.

Photograph 4. Grayish white colored selenite crystals found along Spring Creek.

In contrast, the amber variety had a brighter luster, a better developed "rose" structure, and a single, "flattened", heart shaped monoclinic form usually with minor "petal" development (Photo 5). Commonly, the amber "rose" shaped crystals ranged from 0.5-inch to over 3.5-inches in diameter; however, George Fisher and I found amber "rose" crystals over a foot in diameter, having 1-to-2-inches thick "petals" near their bases, that thinned outward from their centers to less than 0.5-inch thick. The overall length of a "rose petal" was as much as 6-inches. Unfortunately, because of the depth into the bank and the consolidation of the clay host enclosing the crystals, George and I could not extract a complete "rose". So, we filled in the excavation and planned to return in a few days with better digging equipment. Unfortunately, two days later, we found that the Highway Department had sealed off the area to the public to begin construction of the new bridge.

During the highway construction, Bob King, who only lived a few blocks from the selenite site, observed a large backhoe breaking the ground above and near the location of the selenite roses. The back-hoe operator was preparing the ground for the construction of a McDonald's restaurant. Bob, not missing an opportunity, brought a "rose" to the back-hoe operator and asked if he had seen any "rocks" like this. The operator said, "no". But he was curious and asked about the "rose." He then moved his backhoe over to the site above the suspected selenite location described by Bob. The operator dug down as far as he could, but he did not find the selenite zone.

Photograph 5. Amber colored selenite crystals found near the present Fountain Boulevard and Circle Drive intersection.

Years later the McDonald's and the operator's exploration site was demolished to make way for a newer highway.

Prior to this locale being closed and lost forever, the CSMS members as well as myself were able to collect for less than 8 months. Over the years since the overpass construction and west of the original collecting site, smaller (less than an inch in diameter), amber colored "roses" have been found along a portion of Spring Creek. But unfortunately, nothing like the size and abundance of selenite found at the original location.

MANITOU SPRINGS AREA

Crystal Park (El Paso County), located just west of Manitou Springs, produced the best amazonite crystals (Photo 6) in the state (in my opinion). These crystals were translucent robin's egg blue and were as much as two inches long by an inch wide. They occurred with columnar, platy hematite (Photo 7; some individual columns were over 3- inches long), smoky quartz, and green octahedral fluorite (Photo 8). Tan feldspar crystals (as much as 3- inches in length) were also found as single crystals or in groups. Topaz (some green in color) had been reported by earlier collectors, but we never found any during the late 1960's though 1972.

Photograph 6. An uncleaned group of Amazonite from Crystal Park, as found in the field.

Photograph 7. Platy hematite on smoky quartz
from Crystal Park.

Photograph 8. Green octahedral fluorite crystal
(just over a 0.5-inch long) from Crystal Park.

    In late 1968, Bob King, Richard (Dick) Holt, and I began hunting in this area. At that time this locale was very isolated, and I warned Bob and Dick that bobcat and mountain lion roamed the area. Late one summer, Dick invited me to hunt that area, but I was unable to go with him. Dick informed me that he and his dog would go alone. I warned him that his dog might draw in a cat, but Dick ignored my warning and went with his dog anyway. While hunting, Dick's dog was attacked by a mountain lion! Dick just managed to beat off the attack with a rock hammer and a 1940s trench shovel. Later that week,

Dick's adventure was reported in the Gazette Telegraph, a local Colorado Springs' newspaper. Dick's mountain lion adventure is a warning to collectors who search in the Crystal Park area. Also important is that this area is part of a gated private community and at present is not open to the general public.

SAINT PETER'S DOME AREA

South of Crystal Park was a collecting site located on the south slope of St. Peter's Dome. During the 1960s through 1972, the main access to this locale was by a section of the Gold Camp Road between Helen Hunt Falls and the Eureka mine. In the Eureka mine, well-formed, gemmy, red zircon crystals occurred in a bull quartz matrix. The largest zircon crystals that Bob King and I found were about 0.25-inch diameter. Along the slope north of the Eureka mine were numerous pegmatites that commonly contained single (Photo 9) and clusters of brown opaque zircon crystals (less than 0.5-inch long) on tan feldspar or on bull quartz (Photo 10). Also found in the pegmatites were well formed single and groups of white "fish tail" twinned feldspar crystals (Photo 11). The feldspar groups were as much as 4-inches in length and 3-inches wide with individual crystals as long as 2-inches. Associated with the whitish feldspar were well-formed black riebeckite crystals over 2-inches long and about 0.5-inch in width (Photo 12).

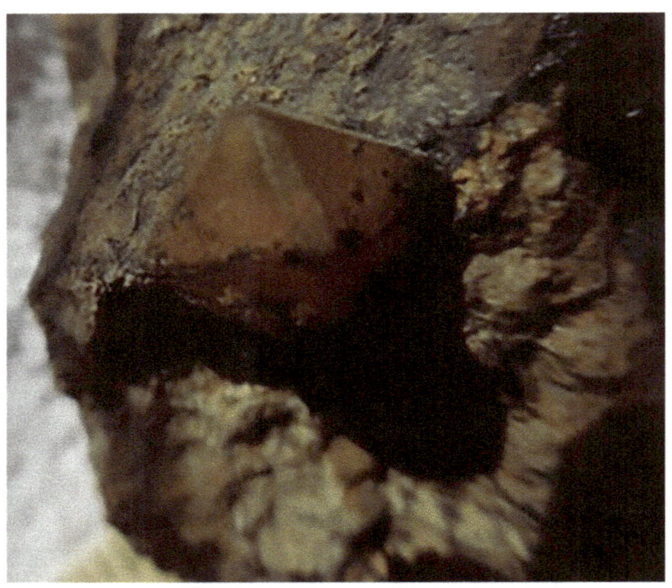

Photograph 9. A tetragonal-dipyramidal zircon
crystal (less than a 0.5-inch long) from
St. Peter's Dome area.

Photograph 10. Cluster and a single crystal of zircons
from Saint Peter's Dome
(Photo from the Andy Kowalski collection).

Photograph 11. Uncleaned feldspar group from Saint Peter's Dome. Note the "fish tail" twinning.

Photograph 12. Riebeckite crystal from
Saint Peter's Dome. Specimen collected by Andy
Kowalski (Photo 2).

Also, found in this area were beautiful gold or bronze-colored astrophyllite that formed clusters of very thin crystals having an elongate habit. These crystals that Bob and I found formed clusters as much as 4-inches long and less than 2-inches wide embedded in/or on bull quartz or on feldspar (Photo 13).

Photograph 13. Astrophyllite cluster in a bull
quartz matrix from Saint Peter's Dome.

Up until the 1960s, I never found a terminated astrophyllite crystal and I never knew anyone who claimed to have found one. However, in the late 1960s, George Fisher was the only collector I knew that had found a terminated astrophyllite crystal cluster (3-to-4-inches long and less 1.5-inches wide). This unlabeled specimen was later sold to a well-known dealer after George and Betty Fisher's death and is now lost.

Although St. Peter's Dome covers a very large area accessible for collecting, there are many claims that require permission to hunt. One must also remember that above Helen Hunt Falls, the Gold Camp Road is closed to 4-wheeled traffic.

FORT CARSON AREA

Another selenite locale was within the Fort Carson Reservation just south of Colorado Springs. These selenite crystals at this locale are straw-yellow colored and were found in mud "rock" exposed along the banks of dry stream beds. These selenite crystals occurred as "blocky", transparent, well-developed, single monoclinic crystals that commonly formed clusters or stacked groups (Photo 14). Single crystals range in size from 0.5-inch to over 2-inches in length, 0.5-to-2.5-inches in width, and as much as 0.5-inch in maximum thickness. Clusters and stacked crystals can be as much as 4-inches in length.

Also found throughout the reservation area, were great examples of cone-in-cone calcite that formed discontinuous beds located just above the gypsum bearing 'mud' rock bed. The individual cone shape (Photo 15),

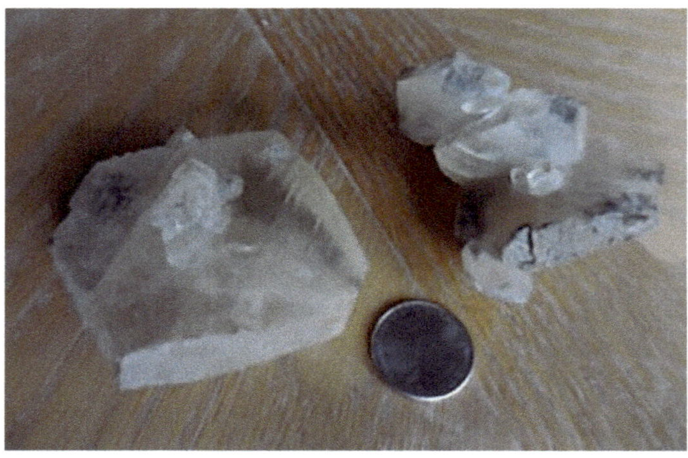

Photograph 14. Selenite crystals from the Fort Carson Reservation.

Photograph 15. Cone-in-cone calcite found on the Fort Carson Reservation.

ranged in length from 3-to-4-inches with an average base diameter of about 2-to-3-inches. The cone shape was so well formed that a past member of the CSMS, Betty Fisher, used them as tepees along with toy "Indians", depicting a native American village that was displayed in one of the early annual Colorado Springs mineral shows.

Once a popular mineral locale for the CSMS, the Fort Carson Reservation is currently closed to the public; however, it may be possible for mineral cubs to obtain permission to hunt in certain parts of the reservation.

CANON CITY AREA

Within the grounds of the Canon City Prison (Fremont County) and at the base of a vertical sandstone formation was a clay bed that produced amber colored selenite "roses" similar to those found in Colorado Springs. However, these crystals were not well formed and were easily cleaved during excavation. The crystal cleavages had a bright luster and were very transparent, resembling glass, with few to no inclusions. A danger at this locale were the scorpions found in and around the selenite occurrence. Ray Ziegler, a member of the CSMC, discovered one of these beasts crawling inside his pant leg. Instinctively he attempted to crush it by slapping his leg, but with every slap he was painfully stung. Nearby "diggers" shouted for him to remove his pants, but because of the females present, Ray refused. Later, Ray counted over 6 stings.

This forgotten locale may still be open to collecting if permission is granted. However, a recent location with similar crystals is reported just east of the High Drive. "Rose" selenite crystals at this site are crude but more complete than those found in the Prison grounds.

FLORENCE AREA

Southeast of Canon City and near Florence (Fremont County), grayish white calcite groups occur along joints, fractures, and fault zones in mud "rock" just above limestone beds. Local collectors described the individual calcite crystal shape as "nail head" calcite (Photo 16). This calcite shape was due to a flattened hexagonal rhombohedral crystal form. The well-formed crystals forming groups were commonly 1-to-2-inches in length and less than 0.5-inch-thick; however, some crystals were over 8-inches in length and 2-to-3-inches wide.

Photograph 16. Typical "nail-head" calcite group found near Florence.

PORTLAND CEMENT QUARRY

South of the town of Florence is the Portland Cement quarry where several limestone beds were mined to produce cement. During the mid-1960's to the early 1970's, collectors were allowed into the quarry where they could find modified cubic pyrite crystals as large as an inch in diameter (Photo 17), elongate groups (1-to-2.5-inches long), and rare octahedral crystals (less than 0.5-inch wide; Photo 18) in the limestone debris. The composition of these crystals was later found to be a mixture of pyrite and marcasite (samples provided by Donley S. Collins for analysis by Pete Modresky of the U.S Geological Survey, personal communication, 1985). It is interesting to note that after 60 years the elongate crystals that I had collected in the 1960s have disintegrated into a grayish white powder. I have also noticed a similar disintegration of "cocks comb" marcasite from the Tri-State mining district. To prevent disintegration, I found that mineral oil sealed and saved these marcasite crystals.

Since the 1970s, the quarry has been enlarged and I'm not aware of any recent reports from collectors as to the status of this locale for collecting or if those "pyrite" crystals can still be found in the newly mined areas.

Photograph 17. Pyrite/marcasite crystals from the Portland Cement Mine quarry south of Florence.

Photograph 18. Octahedral pyrite/marcasite group (about 0.25-inch wide) from the Portland Cement mine south of Florence. This crystal is in the Robert W. King collection.

HOWARD COLORADO

To the west of Canon City is the small town of Howard and a nearby limestone quarry. The limestone from this quarry, once owned by CF&I Company, was used as a flux during steel production. This quarry was also known for cavities containing "dogtooth" calcite crystals. My one and only trip to this locale was in the summer of 1965. Audry Patton, Jack Thompson, Max Filmore with wife Dorothy, and Dorothy's brother Eugene Schachardet, and I arrived one Sunday morning at the entrance to the quarry. Once there, I climbed a nearby ridge and found a "key" shaped calcite group (a foot long and 5-to-6-inches wide) wedged in a vertical "hole". This group consisted of 1-to-3-inches long, opaque, white calcite crystals. After removing the group, I called the others to the site. Using flashlights, we could see about 5-to-6-feet to the bottom of the hole which appeared to be covered with crystals. Because the hole was less than 2-foot wide and being the smallest person in width, I was lowered by hand into the opening. Once at the bottom and sitting on sharp crystals, I found that I was in a cave about 5-foot high and 3-to-4-foot wide, extending more or less north as far as the flashlight could shine. What I saw staggered me. The floor, walls, and ceiling were covered with all sizes and various colored (reddish pink, white, and grayish white) dogtooth calcite crystals 0.5-inch to over 3.5-inches in length. Immediately behind me was a solid wall of massive calcite. I was so stunned that I wasn't aware of the voices asking, "What's going on?" I tried to describe the beauty that was in front of me. Cameras were handed to me and pictures taken, unfortunately none of the photos have been located for this book. I then began very carefully to excavate the calcite groups from the floor to make room to sit and work and soon found that it was not difficult to remove groups over 1-foot long and 4-to-9-inches wide-- a common size. Also, clusters easily peeled off the walls without much difficulty; however, ceiling clusters were somewhat harder to remove and perhaps dangerous. I soon noticed the colors of the calcite crystals varied from a light pink to reddish color on the ceiling, a grayish white (manganese oxide? stain) color on the walls, and a white color on the floor (Photo 19). All crystals had a bright, shiny luster and were portions of scalenohedral (dogtooth) forms.

Photograph 19. Dogtooth calcite from the old CF&I
Quarry northeast of Howard.

Studying the clusters on the floor, I observed that a few ceiling crystals had fallen to the floor where they were later surrounded and overgrown by the floor crystals--perhaps recording a past seismic event or a fluke formed during crystallization. After 2 to 3 hours, our two jeeps were maxed out before I had even started to dig! These crystal groups required very little to no cleaning. I do not know what ever happened to that cave, but years later the property was sold and closed to the public. Today, it is still closed to the public.

THE ROYAL GORGE BRIDGE AREA

Northwest of Canon City is the Mica Lode Mine (Fremont County), mainly a beryl and feldspar quarry. The Mica Lode is located east of and on the north side of the Royal Gorge bridge. In the 1960s, the mine was noted for crudely formed, hexagonal shaped, bluish-green beryl crystals having lengths over a foot and some being greater than 8-inches wide. The two matrix materials were a tannish massive feldspar and a gemmy white bull quartz that made beryl crystal extraction very difficult. As a result, the beryl crystals required a steady hand while using a sledgehammer to remove them from their matrix. One incident of beryl extraction involved two CSMS members, George Fisher and Chris Christenson. George had found a very nice large beryl crystal and was struggling to remove it from the enclosing feldspar. Frustrated after a while, George called Chris (who had a knack for this kind of thing) for help. Studying the problem presented to him, Chris, while holding the specimen, suggested that George position the specimen on a rock and then hit the feldspar matrix along a contact plane bounding the beryl crystal. So, George (while holding the correctly oriented specimen) placed the back of his gloveless hand on a rock and before anyone could stop him--he hit the specimen!!!! Oh, I never witnessed such tears, blood, and moaning...I had to leave. The hammer George used was a painful 2.5-pound sledge; however, the beryl crystal survived the impact!

Also found in the Mica Load Mine were well formed, single and clusters of brownish red almandite (almandine) garnets (Photo 20) mainly associated with the bull quartz. Individual garnet crystals commonly measure less than 1-inch in size and formed clusters measuring 2-to-3-inches in maximum diameter.

Photograph 20. Almandine garnet group from the Mica Load Quarry northeast of the entrance to the Royal Gorge Bridge.

Since the 1970s, the Mica lode has not produced as much collecting material as in the past when it was a favorite field trip location for the CSMS. Currently, I do not know if the mine is open to collecting or if permission can be granted to hunt at this site.

Across from the Mica Lode Mine to the south side of the Royal Gorge and east of the Royal Gorge Bridge is a north facing, white, bull quartz dike having about 8 vertical feet and over 5 feet wide exposure at its base. This dike contained schorl (black tourmaline) crystals as much as 3-inches long and less than 0.5 inch wide. Most crystals had a dull luster with quartz filled fractures (Photo 21).

Photograph 21. Schorl (black tourmaline) in a bull quartz matrix collected from a quartz dike east of the south side of the Royal Gorge bridge.

During the late 1950's and early 1960's, these schorl crystals were very common, but in later years they were difficult to find; a result of over collecting. Presently, this locale is currently claimed, and permission is difficult to obtain.

LAKE GEORGE AREA

West of Colorado Springs and north of Lake George in Teller County is one of the great pegmatite locales in the State. In the mid to the late 1950's, a large portion of this pegmatite locale was once a ranch owned by Ms. Jefferies who would charge $3.00 and later $6.00 per person to collect on her property. The few times that my aunt and I met Ms. Jefferies, she and her daughter were doing their horseback patrol of their ranch area. They were drawn to the noise of our digging or that of other collectors and charged us her fee. Her land, as well as the adjacent Federal lands, were very pegmatite rich, and produced world class quality minerals.

Until the late 1950's to mid-1960's, both Ms. Jefferies and the adjacent Federal lands were far from being over-dug or claimed. It was a perfect area to learn how to locate and dig "pockets" of crystals. Sometime after 1957, Ms. Jefferies' land was sold and developed and soon after, large portions of the adjacent forestry (Federal) lands were claimed by amateur or professional prospectors. This has made collecting locations in the Lake George area scarcer or not accessible. However, during the late 1950's and through most of the 1970's, this was not a problem. My aunt, Bob King, George Fisher and I found that pockets were fairly common and produced many smoky quartz crystals (Photo 22). Other minerals found by Bob and me included salmon colored feldspar (Photo 23), white albite (Photo 24), amazonite (Photo 25), blue fluorite (Photo 26), platy hematite (Photo 27a), rhombohedral hematite (Photo 27b), limonite pseudomorph after hematite (Photo 28a), hematite altered to red ocher (Photo 28b), "onegite" (a local misnomer for amethyst with goethite inclusions; (Photo 29), goethite (Photo 30), zinnwaldite? (Photo 31; later identified by George W. Fisher in about 1995), muscovite (Photo 32), biotite (Photo 33), topaz (Photo 34), and amethyst tipped smoky quartz (Photo 35).

Photograph 22. Example of the many smoky quartz crystals found by Bob King and me in the Lake George area.

Photograph 23. Salmon colored feldspar group from the Lake George, area.

Photograph 24. White Albite with feldspar from the Lake George area. Note size next to portion of the quarter.

Photograph 25. Amazonite group from the Lake George area.

Photograph 26. Cubic, light blue fluorite found in the Lake George area.

Photograph 27a. Platy hematite crystals nested in spray of goethite from the Lake George area.

Photograph 27b. Rhombohedral hematite from the Lake George area.

Photograph 28a. Limonite pseudomorph crystal after magnetite on smoky quartz from the Lake George area.

Photograph 28b. Hematite altering to red ocher on smoky quartz from the Lake George area.

Photograph 29. "Onegite" crystals on goethite from the Lake George area.

Photograph 30. Goethite cluster from the Lake George area. Specimen Photographed from the George Fisher Collection.

Photograph 31. Zinnwaldite (?) the grayish cone shaped crystal on light pink microcline and with white albite.

Photograph 32. Muscovite with smoky quartz from Lake George area.

Photograph 33. Biotite with quartz and amazonite fragments from the Lake George area.

Photograph 34. Very light colored 510 carat topaz from the Lake George area. Photograph and topaz from the Robert W. King collection.

Photograph 35. Amethyst tipped smoky quartz crystals from the Lake George area.

Through the years there have been a number of sites referred to as quartz hill in the Lake George ring dike area. In 1958, my aunt, Betty Westfall (soon to be Mrs. George Fisher) and I made our first trip to Lake George to dig on a hill referred to as Quartz hill at that time. At our first view of this hill, we were stunned by the hill's destruction. Very little ground had escaped a minor or a major excavation. After wandering around the slope of this hill for a few hours, it seemed that the possibility of finding anything was pretty remote. Because this was our first experience at Lake George, we spent most of our time picking through dumps and exposed pegmatite vugs for quartz and feldspar crystals. Later that day, we met Stuart (Stu) Hays, a pleasant man and a member of CSMS. After a short conversation, he found that we had no idea what we were doing. Being patient, he described and showed us how and where the most likely places to dig. Stu stated that most collectors never quite dig out a pocket because of not digging deep enough or starting too far up the slope from the main body of the pocket. He then suggested we try our luck in a nearby digging. JACK POT!! In a short time, we excavated over 25 single-quartz (Photo 36) and feldspar crystals and groups of these crystals. Since that first find, I also found that earlier collectors did not thoroughly explore their packets. For instance, I have found that collectors unknowingly covered the lower portion of the main pocket with their digging waste, or they did not explore further to the right or left of their pockets. As a result of these mistakes, I have been well rewarded with wonderful finds. I have also found that "old" dumps from earlier collectors have revealed many quartz crystals and other minerals that were overlooked or rejected. In rare cases, in one summer, I found as many as two or more rock hammers hidden in dumps. So, never pass by an excavated pocket and its adjacent dump without checking them out.

Photograph 36. My (Donley S. Collins) first quartz crystal from an abandoned pocket shown to me by Stu Hayes. Note the chipped termination, a result of an earlier "digger".

Several years later, in the summer of 1962, my aunt and I returned to Quartz Hill. While searching the lower part of the hill, I found Clarence Coil and his friend Art Reese, both members of CSMS, quietly excavating a pocket. The last crystal that Clarence removed was at least a softball sized champaign colored fluorite crystal. A few weeks later, Clarence invited me to his home to view a nicely cleaved champaign colored fluorite—the same crystal that was recovered earlier. Clarence said that he wanted to show me something special about the crystal. He then pulled out a black lamp and held it to the base of the crystal for what seemed to be about 3 minutes. After turning off the lamp, the fluorite crystal phosphoresced an even yellow color for well over a minute. Being a new collector, I had never seen anything like that. After our conversation about this crystal, he presented me with two cleavage pieces.

Collectors should also be aware of the arrowheads (points) found throughout the Lake George area. Some of the points found by George Fisher in the late 1960s were made of red chert and over 2-inches long. The points that I have found were made of quartz and about 0.5 inch in length (Photo 37).

Photograph 37. One of many arrow heads found in the
Lake George area.

Over the years, I have witnessed some strange
events while hunting for minerals in the Lake George area.
One such event happened in the mid 1960's. During that
time, Bob King and I had invited a friend who was a
geologist and a member of the CSMS to hunt with us. We
had selected the hill where Bob had earlier found a large
topaz (Photo 34). Once at the site, we began moving up
hill while watching for mineral indicators. As I progressed
up the hill, I stopped at a pile of rabbit droppings to see
how old the pile was, a hunter's habit. Afterwards, I
continued climbing up hill. As I proceeded, I turned in time
to see our friend kneel down at that same rabbit pile. To
my astonishment and before I could stop him, he picked
up a rabbit pellet, rolled it between his fingers, and
popped it into his mouth!! He immediately spit it out
while looking up at me. "Smart pills," I said referring to an
old joke. I can't repeat what he had said, but I suggest to

other collectors to be careful what you put into your mouth or what you lick: It may not be a mineral.

As noted, many outstanding minerals have been found during the 1950's and 1960's in the Lake George area and many are documented in reports and museums throughout the world. However, some of the stories behind these finds have seldom been recorded and I will present an interesting one of many that I know.

In the late sixties, a couple stopped at the combination gas station, general store, and museum in Lake George. While looking over the minerals in the museum, the couple asked where the quartz crystals came from. The store owner pointed to the north and said "...just over there in the hills." After spending a few hours and using only a claw hammer, the couple returned with a very large, black-metallic cluster of crystals--so the story goes. The store owner identified the crystal clusters as goethite, having long thin blades with beautiful luster. When asked where the find was made, the couple pointed in the same general direction of the hills that the store owner had pointed out and said, "...to the next hill beyond!!!". The story of the goethite samples (a few given to the store owner) was told over the next several years, inspiring some collectors to search for the open pocket left by that couple. George Fisher was one of those collectors that searched for this locale for at least three digging seasons. One spring, George told me about a dream of a pocket that he found. A pocket that he could almost walk into and that contained a goethite group the size of a loaf of bread!!! For most of that digging season, I listened to his story. Then one day he called and invited me over to his home. He was so excited over the phone and kept saying, "Wait until you see what I have!". When I arrived, to my surprise, was the loaf of bread-sized goethite group just as he had described in his dream. He had found that couple's pocket which produced the largest and finest goethite groups I had ever seen. The terminated blades of some clusters were close to 2-inches long and less than a 0.25-inch wide. During his description of the pocket, I noticed that his right hand was bandaged and asked him about it. George said that when he first reached into the open pocket, which was very dark, he cut his hand on goethite crystals. Some of these groups are now in museums in England and in the Smithsonian. George was generous enough to present me with a small but very fine example similar to the one pictured in Photo 30.

BADGER FLATS

To the west of Lake George is the intersection of HWY 24 and Colorado 77. About 14 miles due North-West of this intersection is Badger Flats (Park County), an area once mined for beryl crystals. The beryl crystals were commonly less than an inch long and 0.25-inch wide. They were brown, well formed, highly fractured and had pinacoid terminations. On closer inspection, some may have been very gemmy before fracturing occurred and some appeared to be aquamarine. Overall, these crystals were once very abundant and occurred in shallow pegmatite pockets. At a nearby occurrence, one could find blue opaque beryl crystals within a white albite matrix. The largest of these beryl crystals observed by me was 3-inches long, a little more than 0.25 inch wide, that graded into a pinkish tan feldspar that maintained the beryl shape--a pseudomorph after beryl (Photo 38).

Photograph 38. Feldspar pseudomorph after a blue beryl crystal. The length of the whole crystal is about 3.5-inches long with a width of less than 0.5-inch. Note the white albite matrix. Specimen found in Badger Flats, north-west of Lake George.

These pseudomorphs were not common at the time, and as far as I can determine, they are unheard of today. To my present knowledge, this locale is probably still open to collecting but may not have the abundance of crystals it once had; a result of over digging. Today and before accessing this locale, be aware of the many claims in this area and always seek permission before collecting.

SPRUCE TREE CAMPGROUND

Just west of Lake George is Colorado 77 that leads north to Spruce Tree Campground (Teller County), once a CSMS rendezvous site for topaz digging. After about a half hour uphill hike to the east of the campground was a location where single white and light to medium blue euhedral topaz, glassy dark colored smoky quartz, and tan feldspar crystals occurred. In the early 1960's, topaz and smoky quartz crystal float were commonly found at this locale, but by the late 1960's, it was very rare to find float. However, in the summer of 1968, float led me to a pocket containing a smoky quartz crystal over 10-inches long and 5-inches wide. Also recovered was about 2-pounds of clear white topaz crystals ranging from 0.25-inch to over 1-inch in length (Photo 39). One topaz fragment, about an inch diameter, was very dark blue in color and when cut, it produced a 5ct stone. Since 1970, the area has been claimed and heavily excavated leaving surface pockets very scarce and floats almost nonexistent. Because this area is claimed, be sure to ask permission to collect.

Photograph 39. Poorly shaped white topaz crystals form the Spruce Tree camp area.

DEVIL'S HEAD

To the North of Spruce Tree Campground is Devil's Head (Douglas County), known to collectors for topaz since the late 1800's. By the time Bob King and I started digging there (from 1965 to 1970), the area had been heavily hunted and unfortunately, we never found a pocket of anything! I chalk this up to over digging and lack of experience. However, during that time, I witnessed others that found beautiful blue and golden topaz associated with some very large (as much as 7-inches long by 3-to-4-inches wide), gemmy, euhedral smoky quartz crystals. The topaz crystals formed masses and beautiful single crystals that were 1-to-3-inches long and 1-to-2-

inches wide. Although this area is still open to collecting, over digging has made surface pegmatites difficult to locate. See Richard M. Pearl (1958, p. 152-156) for a greatly condensed description of this area.

BETWEEN LAKE GEORGE AND THE OLD PIPE SPRINGS CAMPGROUND

A few miles west of Lake George, schorl tourmaline crystals occurred in bull quartz outcrops exposed along the south barrow ditch along HWY-24. These crystals are well formed with some having bull quartz filled fractures. Crystals ranged from 2-to-3-inches long by as much as over 1-inch-wide (Photo 40).

Photograph 40. A terminated schorl tourmaline crystal on bull quartz found in barrow dich along HWY 24. Note the quartz filled fractures. Tourmaline from the Robert W. King collection.

I do not know if those outcrops are still exposed. However, about 13.5 miles west of Lake George is a south turn off from HWY 24 that leads to the South-West to the old Pipe Springs Campground. Near this site in the summer of 1969, Andy Kowalski (Photo 2) discovered schorl tourmaline in a large bull quartz out crop. At this site, Andy found not only schorl tourmaline, as much as 3-inches long with some with quartz filled fractures (Photo 41a), but also almandine garnet, as great as 0.5-inch in diameter, that occurred as single crystals or in combination with tourmaline (Photo 41b and c). Today, this locale is claimed, and permission is required to collect there.

Photograph 41a. Schorl tourmaline crystals in bull quartz found near the old Pipe Springs campground. Note the quartz filled fractures. Minerals from the Andy Kowalski collection.

Photograph 41b. Schorl Tourmaline Crystals from the old Pipe Springs campground. Note the almandine garnet on a tourmaline fragment (center). Minerals from the Andy Kowalski collection.

Photograph 41c. Almandine garnets. Note garnet on schorl tourmaline in upper left of photo. All crystals were collected by Andy Kowalski in 1969 near the old Pipe Springs campground.

HARTSEL AREA

A few miles south of the small town of Hartsel (Park County), located on the western edge of South Park, there a were number of barite prospect pits. One of these pits was noted for straw-yellow barite clusters occurring in a grayish clay. The individual crystals forming the clusters were "spear-point" shaped, but some were more "chisel pointed" or wedge shaped. The barite clusters having the wedge-shaped habit were referred to by early collectors as "cabbage heads". Long time exposure to sunlight changes the straw yellow color of the barite to a moderate or dark blue color. Clusters and single crystals were also found in the adjacent dump.

Clusters from this locale, commonly over 9-inches in maximum dimension, had individual crystals 1-to-4-inches in length, over 2-inches wide (at the base), and less than a 0.5-inch-thick (Photo 42). Some crystals showed well defined zoning (Photo 43).

Photograph 42. Barite clusters from Hartzel. These clusters were not exposed to sunlight (which would turn them blue) and remain a natural light gray and yellowish gray.

Photograph 43. Zoned barite crystal found south of the town of Hartsel.

Unfortunately, barite samples found today lack the quality compared to those found in the late 1950's through 1972. The area is now claimed and requires permission to collect.

La GARITA AREA

Far to the south of Salida is the small town of La Garita (Saguache County). Just north of La Garita and near Biedell Gulch was a volcanic "cone", once mined for gold. But what interested collectors was the mine dump that had single and groups of quartz crystals, having a Muzo habit and some with amethyst "tips" (Photo 44).

Photograph 44. Uncleaned amethyst tipped quartz crystals from the mine dump near Biedell Gulch just north of the town of La Garita. Note: These crystals have the unusual Muzo habit that was characteristic from that dump.

These quartz crystals were very gemmy, commonly less than 4-inches long, and generally contained water bubbles. In the early days, miners would dare each other to drink the water from these bubbles, believing them to contain cyanide. However, in about 1965, Professor Richard M. Pearl at Colorado College in Colorado Springs investigated this idea. He contacted Chris Christenson, a member of the CSMS, and asked if he could provide quartz crystals having bubble inclusions for analysis. Chris, being very honored by this request, provided Professor Pearl with three of his very best amethyst tipped quartz crystals, each over 2-inches in length, and each having large water bubble inclusions. Professor Pearl was very pleased with these samples and promised to return them. After three months, Chris visited Professor Pearl and asked how his study was progressing. Pearl happily stated that the study was completed and that the water inclusions did not contain cyanide. Chris congratulated Pearl and asked if he could have his crystals back and Pearl promptly returned them- but in pieces!!!! which immediately generated the question "WHY?" from Chris. To which Pearl answered, "...that was the only way to analysis the water was to extract it (the water) from the quartz samples!" Later in the 1970s, the water contained in these crystals was used to determine the temperature and chemistry of mineral formation.

Commonly coating the quartz crystals and most of the dump material was a very dark gray colored, sooty wad--a manganese oxide mixture--that smeared easily, ruined clothing, worked its way into your skin, and made digging in that dump a very dirty experience. The wad also formed individual dendritic columns (Photo 45) and in combination with quartz crystals (Photo 46).

Photograph 45. Sooty dendritic wad columns with imbedded quartz crystal from the mine dump near Biedell Gulch north of La Garita.

Photograph 46. Wad (dark color) on quartz crystals found in a mine dump near Biedell Gulch north of La Garita.

In the mid-1970s, a fellow graduate student from CSU investigated this site and found that the wad contained a high content of gold, indicating that this mine had a good potential for further development. A mining company, based on the students' findings, drilled this site and found a layer of gold over 1-inch-thick in the core. The company later excavated the subsurface gold deposit and processed it and the dump material containing the wad and the quartz crystals. The dump and in situ gold- bearing material was crushed and cyanide-leached to recover the gold.

Before mining started, large boulders found at the base of the dump contained groups of white Aragonite crystals. (Photo 47). The individual aragonite crystals forming the groups were less than 0.25-inch in maximum dimension and fluoresced white and bright yellow-green color under short wave light. Unfortunately, this material was also destroyed during the mining operation in the late 1970s and early 1980s. Thus, another great mineral locale was lost. Also, because this site is on mine property and as a result of the cyanide leaching process, I believe this site is currently closed to the public.

Photograph 47. Aragonite crystals covering dendritic wad "crystals" from a mining dump near Biedell Gulch north of La Garita.

SALIDA & NATHROP AREA
SEDALIA COPPER MINE

North of Salida (Chaffee County) is the Sedalia Copper Mine well known for euhedral, dodecahedron almandine garnets that formed in a green micaceous schist. The schist exposure is located just over 400 feet into the north trending mine tunnel in an intersecting east trending adit. Forming the south facing wall of this east adit is green micaceous-garnet schist, containing the garnet crystals. During the 1960s, Bob King and I extracted many garnets (Photo 48a) from the schist that ranged in size from a few inches to more than 5-inches or more in maximum diameter. Clusters of two or three garnets were rare, but single crystals were common (Photo 48b).

Photograph 48a. Bob King "digging" out garnets in the Sedalia Copper Mine in 1966.

Photograph 48b. Almandine garnet crystals from the Sedalia Copper mine.

Once extracted from the rock, one had to carve the garnet surfaces to remove the remaining schist; thus, exposing the sharp garnet form. The garnet surface, after carving, showed a mixture of garnet and schist. Wondering how deep the schist penetrated the garnet, I broke a few garnets and found that the schist penetrated deep into the garnet crystal structure. However, I also found that some garnet centers to be schist free, unfractured, and having very gemmy dark red color.

Further north within the main Sedalia Mine tunnel, clusters of blue needles of chalcanthite crystals were found on the tunnel walls and support beams. The clusters were from 1-to-2-inches across with needles as much as 2-inches long. Being very delicate, these clusters were very difficult to collect as well as to preserve. While searching near a sheared beam in the late 1960's, George Fisher and I found several great clusters of chalcanthite. One sample, found by George, consisted of two separate clusters: one on the upper and the other on the lower surface within a wedge-shaped matrix. Both clusters were about an inch wide having radiating 0.5- inch chalcanthite needles, a great specimen! A few weeks later, during a visit to George's home, George's wife Betty insisted that I take a closer look at that cluster. Because it was a very fragile specimen, I suggested that she not bring it out of the display case, but she did. While she held the specimen at eye level, the wedge collapsed!!! A beautiful specimen lost. Betty always seemed to have a problem with holding minerals. For example, in 1965 at a CSMS mineral show, a well-known dealer allowed Betty to carefully hold a 3-inch sized cross shaped wulfenite crystal from the Red Cloud mine, Arizona. As George, the dealer, and I were standing around, Betty had the crystal suddenly lifted from her hand and it "flew" to the floor! We were shocked at the sight. How did this happen! There was no breeze, no one bumped her, and she had not moved; the sample just flew out of her hand! From that experience with Betty, I never allowed her to handle any of my minerals.

Presently, the Sedalia Copper mine is owned and permission to collect is difficult to obtain. If you are one of the lucky ones to gain permission, be very careful if you try to collect the chalcanthite crystals. In that part of the main tunnel, where these crystals occurred, appeared to be unstable. This was suggested by a sheared "tree trunk" support beam over 10 feet long and about 5 feet wide and wedged at an angle of about 25 degrees between the tunnel walls.

CALUMET IRON MINE

A few miles south of the Sedalia Copper mine is the Calumet iron mine (Chaffee County) that is in the Turret district. In the 1950s through 1972, a steep path (located on the south side of the mine dump) led upwards to the mine entrance. Most people would search the dump for epidote, magnetite, and uralite (a pale green hornblende altered from pyroxene). Many different minerals recovered from that dump were impressive, but one mineral, an epidote crystal, had an interesting story. This epidote crystal was found in two separate parts on two separate occasions, each part measuring just over 3-inches in maximum length. The first part was found in the Calumet Mine dump by George W. Fisher, and the other part was found about a year later by Mrs. Mary Rapjack, a member of CSMS, in the dry stream bed near the base of the dump. Comparing the two crystal parts at a CSMS brag session, both owners claimed to have the largest epidote crystal ever found from the Calumet Mine. However, on close inspection, George Fisher found that both parts fit perfectly together. So, George and Mary promised that if anything happened to either person, then their part of the crystal would be donated to the surviving party. Sadly, George passed away first and his collection was sold to a well-known dealer; thus, George's half of the crystal (not labeled) was lost and years later Mary's half was also lost.

In 1963 (my first trip to the Calumet Mine), Max Fillmore led me to a weathered zone just above the mine opening. The zone was defined by a white, dusty, unconsolidated clay that yielded rare epidote groups. (Photo 50) and single epidote and quartz crystals. Epidote groups were less than 5-inches long and 2-to-3-inches wide and had individual dark green epidote crystals less than 0.5-inch. The single high lustered, dark green, epidote crystals, commonly less than 2-inches in maximum dimension, typically had partial terminations with only two sides ("prisms"; Photo 51). Complete single crystals were almost impossible to find (Photo 52). Also, found in the weathered zone were single, very clear quartz crystals (0.5-to-3-inches long) and commonly having inclusions of masses or needles of epidote. In

1969, a seam containing groups of clear quartz crystals was found by Bob King above the mine (Photo 53). Both single and groups of quartz crystals in pristine condition were recovered and required very little cleaning.

Photograph 50. Epidote group from the Calumet Mine.

Photograph 51. Partial epidote crystals found above the Calumet Mine. Specimen from the Andy Kowalski collection.

Photograph 52. Rare, complete double terminated epidote crystal found in the weathered zone above the Calumet Mine.

Photograph 53. Clear quartz crystal from just above the Calumet Iron mine. Note the epidote inclusions. This photo and crystal are in the Robert W. King collection.

At the top of the mine entrance and north of the weathered zone was an outcrop of massive, fibrous tremolite/actinolite (Photo 54), having a high luster and colors that included white, reddish brown, and green. In 1965, Max Fillmore led me along a very narrow and very slippery trail to that outcrop at the base of a 5-foot pine tree. Once at the site, we had to wedge ourselves against that tree in order to take turns to dig. One slip and we would have fallen a great distance, but at the time it seemed worth the risk. Today, I believe that the outcrop is still there and accessible; however, I would highly recommend to anyone trying to find that site to be very careful. It's a very long fall!

Photograph 54. Fibrous tremolite/actinolite collected above and north of the Calumet Mine entrance.

I never hunted on the Calumet mine dump but always spent most of my time searching the weathered zone above the mine for minerals. However, in 1968, Bob King, a member of CSMS, and I first descended into the Calumet mine by a 40-foot ladder that was missing the lower 2 to 3 rungs! Once at the bottom of the ladder, we found that light from the above mine entrance provided very little to no visibility within the mine. When we lit up the mine, we saw a room over 100 feet in diameter with a celling well over 70 feet high. In the center of the room was a major "hole" or shaft over 60 feet in diameter and having a depth beyond the reach of the light from our flashlights. Along the north wall bounding this "hole" was a pile of calcite and other debris. Searching the pile, we found uralite and magnetite crystals embedded in the massive calcite "chunks." Using Muriatic Acid to remove the surrounding calcite yielded uralite groups as much as 4-to-5-inches in length, as much as 3-to-4-inches wide, with individual crystals over 3-inches long and 0.5-inch wide (Photo 55) and groups (2-to-3-inches in maximum diameter) of shiny octahedral magnetite crystals (Photo 56).

Photograph 55. Uralite group collected by Andy Kowalski within the Calumet Iron mine.

Photograph 56. Magnetite crystals found within the
Calumet mine. Note the white partially dissolved
calcite matrix.

Individual magnetite crystals forming the groups were commonly as much as 0.5-inch in maximum size. In order to preserve the magnetite's shiny luster, Bob King found that a very weak concentration of Muriatic Acid could be used to expose them from the calcite; too strong an acid concentration and the shiny luster was lost.

Today, like so many areas, the Calumet Mine is closed, and one must acquire permission to hunt in this area.

RUBY MOUNTAIN AREA

North of Salida and east of the town of Nathrop (Chaffee County) is a "group" of three volcanic rhyolite outcrops: Sugarloaf Mountain, the largest and furthest north, Dorothy Hill, the smallest outcrop, located west of Sugarloaf, and Ruby Mountain south of Sugarloaf. Of these three outcrops, the southern end of Ruby Mountain produced the largest (as much as 0.25-inch in maximum size) garnets during the late 1950s through early 1960s. The garnets (variety: spessartine) are transparent, very gemmy, and have a dark red color (Photo 57). The largest garnets (a little less than a 0.25-inch) were found by Bob King and me on the south facing flank of Ruby Mountain in the late 1960s.

Photograph 57. Spessartine garnet crystals on rhyolite matrix from Ruby Mountain.

Other minerals found at Ruby Mountain included transparent and golden topaz, generally less than 0.5-inch in length and less than 0.25-inch wide, and black shinny hematite needles less than 0.125-inch long.

Sparsely covering the ground between Sugarloaf Mountain and Dorothy Hill were well rounded, black, gemmy Apache tears (obsidian) that were commonly found in the late 1950's and early 1960's; however, after the mid-1960's, the Apache tear drops became scarce to find as a result of over collecting. But in 2014, a runoff channel near Sugarloaf produced hundreds of small, less than 0.25-inch in diameter, Apache tear drops.

Since 2015, unfortunately, part of the Ruby Mountain locale is now part of a National Monument, and the other part is privately owned and off limits to the public. So, I would strongly advise not to collect in the Ruby Mountain area unless you can gain permission to hunt on privately owned land.

MOUNT ANTERO & WHITE MOUNTAIN

North of Salida and west of Nathrop (Chaffee County) are the famous mountains Antero and White, both known for outstanding crystals of aquamarine-- Colorado's State gemstone. These crystals are gem quality and range in lengths from less than 0.125-inch to well over 1-foot, having widths of less than 0.125-inch to well over 2-inches. Aquamarine (aqua) crystal habits include hexagonal-pinacoid, Hexagonal-dipyramidal-pinacoid, and dihexagonal-dipyramidal-pinacoid forms. Other minerals commonly found include smoky quartz, topaz, fluorite, feldspar, and phenacite.

My first trip up Mount Antero was in the summer of 1964. It was an amazing ride that ended on a knoll near the trail that led to the hiking register on the summit of

Mount Antero. As I stepped out of the jeep, I found my first golden topaz, about the size of a quarter. Excited, I began to dig and soon found single and twinned feldspar crystals, all having a white glazed porcelain texture. The twins included Carlsbad and Baveno (Photo 58) forms, having lengths of 3-inches or less. Also found were very black smoky quartz crystals that tapered toward their terminations (Photo 59) and have distinctive three sides of high luster and three sides of dull luster. At the end of the day, I buried that unfinished pocket and never returned.

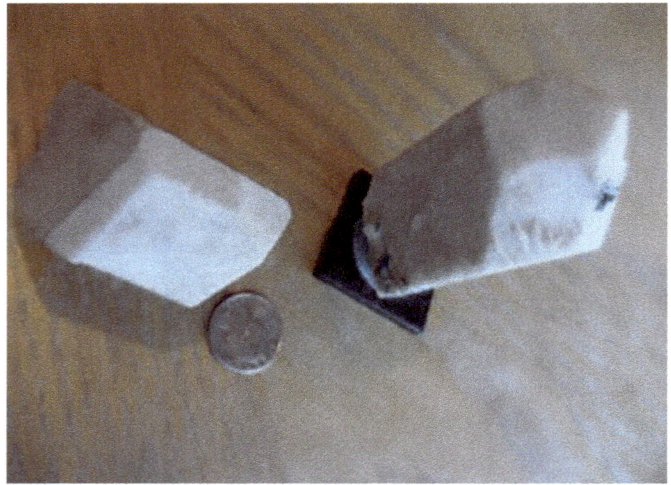

Photograph 58. Oblique view showing the terminations of white, Baveno twinned microcline from Mt. Antero. Note the porcelain like luster.

Photograph 59. "Tapering" smoky quartz crystals from Mt. Antero.

In 1969, George and Betty Fisher, Audry Patton, Jack Thompson, and I were led to the California Molybdenum Mine, just west of Mount Antero, by George Robinson, a well-known collector at the time. Entrance into the mine was a north trending tunnel with mine cart tracks. Just a few yards into the tunnel is an east trending adit that led to an overhead platform. Using a ladder to gain access to the top of the platform revealed an overhead vertical stope just wide enough for two people to work at a time. Once wedged into the stope, several feet above the platform, it required two people, one to dig and another to hold the flashlight and catch the falling crystals. George and I were the first pair to work the stope, and we found it very difficult to brace ourselves between the stope wall and at the same time dig. Every 20 minutes the men would rotate and at the end of the day we would share our findings. What we divided were clear (goshenite) to dark blue hexagonal-percoid aquamarine crystals as much as 3-inches long (Photo 60), purple octahedral fluorite (rare, less than an inch long), and very rare and soft molybdenum hexagonal crystals.

Photograph 60. Donley Collins' share of aquamarine crystals from the California mine just west of Mt. Antero.

The largest molybdenum crystal that I found was a single (about 2-inches in maximum diameter and an inch high), greasy, very soft, silver-colored crystal having a hexagonal-pinacoid habit. Because this mineral was very soft, it was extremely difficult to remove from the stope walls.

South of Mount Antero is White Mountain, a mountain that fascinated me during the 1960s by the rumors and factual stories of the treasures found on its slopes. The earliest story of Mt. Antero and White Mountain told to me was about Ed Over, a well-known collector and co-founder of CSMS. It was rumored that one summer, he had found so many quartz and other crystals that he had to stash them somewhere along the trail leading up to White and Antero mountains. Years later when I descended from White Mountain into the Browns Creek valley I found a cabin near a beaver dam at the head of the Browns Creek. Studying this old cabin, I wonder if Ed stayed there and if one of his many reported mineral stashes could be still buried nearby.

A factual story told about White Mountain began with a very liked member of CSMS. He was an engineer and well known as a very successful mineral collector. In the 1960's, an article in a mineral journal that described aquamarine collecting on White and Antero mountains and included a full picture of White Mountain. Using that picture of White Mountain and a magnifying glass, the engineer located a trend of excavated sites. With this information, he later found a spectacular "pocket" that produced dark blue aquamarine (aqua) crystals with some in combination with smoky quartz. His find is rumored to have filled more than 5 beer flats. It wasn't until 1967 that my aunt and I tried our luck on White Mountain and in the same general area found by the engineer. It wasn't until late afternoon that my aunt found the beginning of a great "pocket" that produced hundreds of "pencil lead" sized (2mm in diameter and as much as 0.5-inch long) aquas. Digging deeper, she found an 8-inch long by 0.5-inch wide terminated aqua, a 4-inch long by 3-inch-wide aqua crystal fragment, and a very dark blue 3-inch long by

1-inch wide aqua. All terminated crystals were dihexagonal-dipyramidal-pinacoid forms. Because it was very late in the day and our ride wanted to leave before it became too dark, we buried the "pocket". Returning early the next weekend, my aunt discovered that the "pocket" had been dug out! When you find a pocket as rare as my aunt's, never show or brag about what you have found. Unfortunately, if you do, you could lose that great find as a result of someone taking it from you. Even worse, it could be a supposed "friend." I have found over the years that this has always been a problem that collectors face. A few months after my aunt found her empty "pocket", I found a potential site near her lost "pocket". It was producing 1-to-2-inch aquas, but about 1.5-feet down, the seam was completely cemented in clear crystalline ice! Suspended about mid-way into the iced seam were 2-to-3-inch long aquas. Running out of time and because the width of the seam created a very small workspace, it became impossible to remove the aqua crystals, so I buried that seam. Years later I viewed that site from Mount Antero and was shocked by the destruction, but I had to laugh because whoever made that very long trench just missed and had covered my ice-covered aquas with the trenching waste.

In the summer of 1970, I made my last digging trip with Bob King and Dick Holt to White Mountain. While searching White Mountain's slope, I found a cave(?) in a vertical granite wall. Calling the others over, we began digging and soon were rewarded with many sizes of smoky quartz crystals (Photo 61). One of the longest quartz crystals found (well over a foot long) was recovered by Bob King (Photo 62).

Photograph 61. Some of the smoky quartz crystals recovered from a cave on White Mountain.

Photograph 62. Bob King with the largest smoky quartz crystals found in the cave on White Mountain.

Fluorite found on White Mountain commonly forms deep purple octahedral crystals. In 1968, Richard M. Pearl showed Donley S. Collins a perfect purple octahedral fluorite (from White Mountain) over 3-inches long and about 2.5-inches wide.

Other minerals found on White Mountain included white (some gemmy) phenacite, white microcline (some Baveno twinned), and an unknown mineral. The phenacite crystals (less than 0.5-inch in size) were generally found coating white feldspar. The dark colored unknown mineral, found at the "base" of White Mountain, occurred in white bull quartz and was difficult to expose or remove. These dark colored minerals were as much as 3mm wide and appear to form octahedral crystals (Photo 63).

Photograph 63. Unknown, dark colored metallic, octahedral shaped crystal in a quartz matrix. The crystal is about 3mm high and found on White Mountain.

Golden topaz as large as 3-inches long were often found on Mount Antero, but were not found on White Mountain by Bob or me during the 1960's.

A warning to anyone who attempts to hunt on Mount Antero or White Mountain. Altitude is always a problem for some, but weather can also be a problem as well. George Fisher found this out while digging on White Mountain. Excited about what he was finding, George forgot to watch the clouds accumulating overhead. Soon he felt like bugs were crawling over him and realized that an electrical storm was overhead. Looking around, George froze as he saw a basketball-sized bluish ball of lightning, bouncing over the boulders not far from him. George said that he started throwing his belt, coins, tools, and anything made of metal away and ducked into lowest spot on the ground that he could find. So be careful and always watch the clouds. Today, the majority of the Mount Antero and White Mountain and surrounding area is under claim and access to collect can be very difficult to obtain.

TEXAS CREEK

Northwest of Mount Antero was a working feldspar quarry located just north of Texas Creek (Chaffee County). Weekend trips to this site were allowed in the quarry in the mid-1960's. After blasting during the week, one could find 1-foot long by 5-inches wide or smaller Baveno twinned microcline twinned crystals (Photo 64a and 64b).

Photograph 64a. A large Baveno twinned microcline crystal from a feldspar quarry north of Texas Creek.

Photograph 64b. Microcline from a Texas Creek quarry. Note the Baveno twining and the albite bands called perthite, very characteristic of microcline crystals from this quarry.

My first and only visit to the Feldspar quarry was with members of CSMS. Once at the quarry, I bolted from the jeep and was first to the blasted quarry wall and found the best crystals (Photos 64a & 64b). The others were surprised how fast I was and suggested that next time they would tie lead weights to my legs.

This quarry is currently inactive, and I don't know if one can access this site any longer. However, in nearby Texas Creek, one could find flat oval shaped, white micaceous schist pebbles. These pebbles were very sparkly, a result the muscovite content, and were commonly 3-to-4-inches long in maximum length with a maximum thickness of about 0.5-inch.

LEADVILLE AREA

About 50 miles to the north of the Calumet Copper Mine is Leadville (Lake County), a historic silver mining town that was later known (during the 1950's through 1970) to collectors for the pyrite (Photo 65) rich dumps throughout the mining area.

Photograph 65. Pyrite group from the Gallager Ghost Mine (Just east of the Little Johnny mine), Leadville mining district.

In the late 1950's to about the late 1960's, the majority of the main business buildings in Leadville were original (dating from the late 1880s) and often closed. When I first visited this town at the age of 10, some of the old buildings were open for business. At that time, it was not quite a ghost town and entering some of buildings open to the public was like stepping back into time. East of town I saw many mine dumps, but the most popular dumps containing abundant pyrite were those of the Ibex Mine(s). The Ibex dumps were over 60 feet high and had very steep (30 to almost 40-degree) slopes. These steep slopes were maintained by a 1-to-3-inch-thick crust formed from a mixture of sulfur and sericite clay; thus, making climbing and digging very difficult. The best pyrite zones within the dumps were determined by the exposed wood fragments from the old mine car tracks.

In order to reach a pyrite zone in the Ibex dumps, deep steps and a bench had to be cut through the hard crust slope. If the steps or bench were not carefully constructed, one could experience a very long and painful fall. Once the crust was penetrated, one was exposed to unconsolidated sulfur and sericite dust that stained and saturated clothes, stained and dried hands, and covered one with a very strong sulfur smell. For a 10-year-old at the time, hunting on the Ibex Mine dumps was a very challenging adventure. During this time, I must credit Clarence Coil, who patiently taught me how to locate, assault, and find pyrite in the Ibex Mine dumps.

The pyrite zones within the Ibex Mine dumps yielded many complete and portions of cubic pyrite crystals with some being as much as 4-inches across. But what mainly attracted collectors to these dumps was the "cathedral" pattern development (Photo 64) on some of the crystal faces.

Photograph 66. Pyrite with the distinctive "Cathedral" pattern was quite unique and characteristic to the Ibex Mine dumps.

Because the Ibex dumps contained abundant pyrite, most were, in the 1960's, hauled away for sulfuric acid production. The few Ibex dumps remaining had very scarce pyrite. Today, these dumps, as well as others in the area, are owned and require permission and sometimes a fee to gain access to dig on them. Another problem is the housing and hiking trail development extending east of town and into the mine dump areas, a future problem for both collectors and investors.

In the general area of Leadville, dendrites occur on surfaces of many rocks and rock fragments. They were formed by crystallization of a foreign material, usually from a manganese solution. These branching deposits are often identified as plant fossils. In the late 1950's through 1970, great examples of well formed, dendrites could be found near old "String Town", now a suburb of southern Leadville.

Photograph 67. "String Town" dendrites.

Between Leadville and the Climax Molybdenum Mine is an exposure of Lincoln Porphyry located on an old, abandoned section of State Highway 91, just west of the Arkansas River that parallels the LC&S Railroad. Between the present highway and the old road section was a steep slope of exposed Lincoln Porphyry. Weathered out of this rock slope were very abundant single and double-terminated, Carlsbad twinned orthoclase crystals, having maximum lengths of about 1-inch and about 0.25-inch-

Photograph 68. Single and Carlsbad twinned orthoclase crystals. Note the porphyry matrix with an orthoclase phenocryst.

wide (Photo 68). They were very glossy, had inclusions, and were very popular to collect during the mid-1960's.

After collecting pyrite in the Leadville mine dumps, members of CSMS would often stop at the old HWY 91 site to collect those weathered out orthoclase crystals by the gallon bucket full. In 1970, I stopped at this locale to collect, but to my surprise, I could only find a half dozen crystals. Although a few in situ orthoclase crystals were present, the weathered concentration of these crystals on the old roadside slope was almost completely "picked clean"; sadly, it is a result of over collecting.

An interesting story about a different orthoclase locale was told, in 1966, to Bob King during a CSMS mineral show by the city fire marshal who had lived in Leadville. During the conversation, the marshal described some "rocks" he had found north Leadville and the next day he brought an old cigar box containing his "rocks" for Bob to identify. The "rocks" were orthoclase crystals, some being 4-inches long. Happy about the identification, the fire marshal gave the location for those crystals and later that summer, Bob and I drove to Leadville to search for that orthoclase locale. While driving along main street through Leadville, we spotted and stopped at a very dusty, dark, "antique" shop. The owner of the shop was a crusty-looking fellow who hadn't shaved for a hundred years. While trying to see though very dusty 19th century glass cases, a well-dressed lady came in, looked around then asked, "…is that item for sale?" Upon which, the owner answered, "Lady the whole joint is for sale!". At the end of this amusing conversation, I noticed two very large 3-inch long by 0.5-inch-wide orthoclase crystals. Bob asked where they were from and was directed to an outcrop just north of town. After comparing the marshal's description to that of the shopkeeper's, we concluded they were for the same orthoclase site. Later that day, we found more orthoclase crystals in or out of their matrix than we could carry. The crystals were 4-inches or less long and had monoclinic forms that included single and twinned crystals. The twins found were Carlsbad (most common), and Baveno (less common). Many interesting combinations of these forms were also found throughout this site (Photo 69).

Photograph 69. Orthoclase crystals collected from the Lincoln Porphyry near Leadville. Note the different twin types and combinations of crystal forms.

Also found in the porphyry were quartzoids, a hexagonal dipyramid form (Photo 70). These uncommon euhedral crystals were crude, dirty white colored, and generally less than a 0.5-inch in maximum length. This locality and surrounding area, where not owned or claimed, are still open to collecting. In general, the Lincoln Porphyry crops out over many square miles offering collectors the opportunity of finding many orthoclase and associated quartzoid crystals. However, when in doubt about access, always make sure to seek permission before collecting.

Photograph 70. Quartzoid quartz crystals from the Lincoln Porphyry north of Leadville.

THE GILMAN AREA

Northwest of Leadville is Gilman (Eagle County) and the nearby Eagle Gold Mine. In the late 1950's through 1960's, minerals recovered from the mine included pyrite (Photo 71), golden barite (Photo 72), rhodochrosite (Photo 73), marmatite (a blackjack—a black colored sphalerite), siderite (Photo 74), and very rare light green apatite.

Photograph 71. Pyrite from the Eagle Mine.

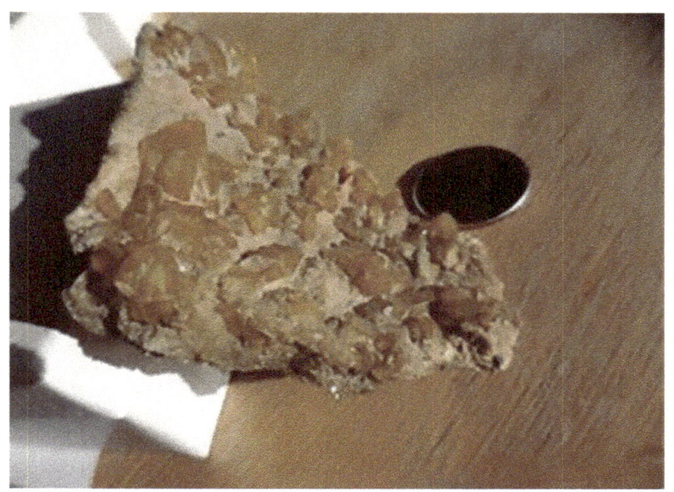

Photograph 72. Golden barite is typical of the Eagle Mine.

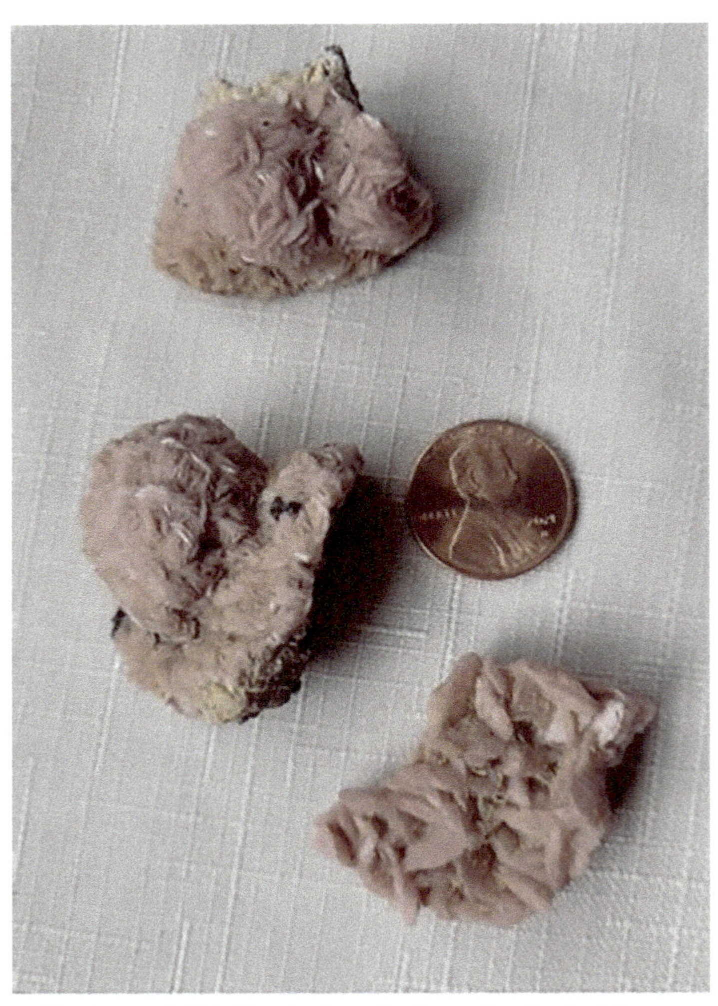

Photograph 73. Rhodochrosite from the Eagle Mine.

Photograph 74. Siderite group from the Eagle Mine.

During the 1950's through the early 1960's, collectors were not allowed on the Eagle mining property, including the dump, but we could buy minerals from the local miners and their children. It wasn't until 1964 that my aunt and I accompanied Max and Dorothy Fillmore to a bar in Red Cliff, a few miles southeast of Gilman. When we entered the bar, I was told that the local miners would sell or trade minerals for drinks. After the bar closed, the bar owner would box and store those minerals in his dark, dirty cellar. Once in the cellar, we found beer flats of abundant pyrite groups which the bar owner sold for $20 a flat—your choice of flats. The pyrite groups were less than 7-inches-long and 4-to-5-inches-wide, having individual pyrite crystals less than 2-inches in size. Pyrite and other minerals of better quality sold for around $20 apiece. Pyrite groups that have minerals such as marmatite, siderite, and or golden barite were common. Rare siderite groups over 5-inches in maximum diameter, having individual siderite crystals as much as 0.5-inch wide, were sold for $5 to $20. Gemmy golden barite groups, 3-to-4-inches wide were priced at $8 to $15 and Marmatite groups, depending on specimen size, varied from $6 to $10 apiece. Most interesting samples sold were very rare groups of marmatite having yellowish-green apatite crystals, some 0.5-inch long and a little less than 0.5-inch-wide.

Today the Eagle Gold Mine is closed and flooded; however, if one searches the mineral shows and shops, these great mineral samples from this mine can still be added to a collection.

STONEHAM AREA

Southeast of the small town of Stoneham (Weld County), Colorado was one of the most famous Barite locales in the United States. At this locale is an altered, very consolidated volcanic ash that is host to very well terminated, dark to light blue, "coffin" shaped barite crystals that often formed with yellow calcite (Photograph 75).

Photograph 75. Blue Barite from Stoneham. Note the whitish-yellow colored calcite on the barite group (right).

My first visit to the Stoneham area was on a CSMS field trip in 1961. Once at the collecting area, Max Fillmore, my ride, led me to the south side of a hill to a weathered seam where we found many single barite crystals as much as 4-inches in length. My last visit was with Bob King to that same seam in 1968. At the end of the day, we collected many single and rare groups, as much as 4-inches in maximum length of blue barite (Photo 73). Before we left, I concluded that the barite seam was along either a north trending joint or fault.

Earlier in the late 1950's, Clarence Coil and Art Reese (past members of the CSMS) excavated a west-east tunnel somewhere on that same hill that Max, Bob and I later collected on. After digging 26 plus feet, they intersected a fault(?) with an open seam of unaltered barite groups that were coated with yellow calcite. The groups were so abundant that they had to leave some in the tunnel which they later buried. They never returned

105

for those groups (Clarence Coil, personal communication, 1965). No one to my knowledge has ever located that tunnel.

One last story about the Stoneham barite happened to me in 1962. Being a poor 12-year-old, I lacked the funds to buy one of Coil's barite groups. However, after searching the ads in a popular journal titled "Rocks and Minerals," I found that Emo Brandenburg, a nearby neighbor, was selling barite groups from Stoneham, Colorado at a modest price. When I visited Mr. Brandenburg, he produced several beer flats of barite groups from which I selected one of the best (a five-dollar 3-by-4-inch group) for my aunt's birthday present. She was quite happy with the present, but upon close examination, she asked me to take a close look at the barite crystals forming the group. I found that a number of the barite terminations were buried into the matrix, where others were oriented correctly above the matrix. With some tests, I found that the matrix was a mixture of glue, clay, and barite crystal fragments. It was a classic Emo Brandenburg creation, one that my aunt cherished with fond memories of this interesting and very kind man.

Unfortunately, today, that Stoneham blue barite locale has been claimed and is no longer open to collecting. However, the volcanic ash bed is very extensive, and one may find an unclaimed source for blue barite. I also believe with the present seismic technology; one would be able to locate barite bearing fractures or faults in the area.

SIVERTON/OURAY AREA

In southwestern Colorado are the mining communities of Silverton and Ouray (San Jaun County). At the end of the Silverton-Ouray train line during the late 1950's and through the 1960's, the children of the local miners would have beer flats full of minerals to sell to the tourists. Some of these samples were really great and included calcite coated quartz (Photo 76), rhodochrosite, fluorite, white calcite that fluoresced pink, and a few sulfides. However, the white opaque quartz crystals forming groups and single crystals coated with druse calcite were the most popular with collectors and tourists.

Unfortunately, the mines are now closed, and the children have moved on with their families, thus making collecting from this locale mainly restricted through dealers. For instance, there are a few mineral shops in both Ouray and Silverton as well as in Durango that may have some specimens available from the Ouray/Silverton area, but mineral shows are the best source for these minerals.

Photograph 76. White calcite coated quartz from the Ouray mining district.

SUMMARY

Although some mineral localities, as described in this document, may be depleted or closed to collecting, one may still find a desired mineral from a lost locale from web sites, mineral shows, other collectors, or even museums. However, beware of mislabeled minerals and/or locales. Even museums sometimes have mislabeled minerals and/or locales. For example, I discovered in a college museum a very large (one foot by 10-inch wide in maximum dimension) group of "rose" shaped rhodochrosite crystals from the Eagle Mine labeled as "pink selenite from Mexico". Also, it is not uncommon in collections to have minerals lacking a name and location labels. This lack of proper labeling causes not only the collector, but also dealers, to guess at that mineral's true name and/ or location.

Also described in this document is the continued removal of collecting areas due to local development, an increase in claims, private land restrictions, stricter public and federal mining rules, and regulations that are continuing to affect present and future collecting. Another contributing factor that has inhibited collecting is the removal of mine dumps and road access to early collecting areas. If you intend to visit any of the above collecting sites or sites not described in this document, be sure you follow all regulations required by the law and always seek permission.

AFTER THOUGHTS FOR FUTURE PROSPECTORS

Colorado has many known and yet to be discovered mineral locales. To find these sites one must know the host rock that the desired mineral or minerals can be found. These sites are found by much hunting, experience, and patience. To aid this experience, one can narrow the search for minerals by using geologic maps that may show the distribution of the mineral host rock. Also, find old and new mineral literature and other collectors' information or attend a mineral show to get information about new sites.

The literature that I have found most useful are the old and current US Geological Survey publications and current mineral journals. These resources have led me to many new and unexplored mineral sites in the past. One thing you must remember, if you find a new site be careful who you share this information with. Good Hunting!!

School Photo ID of me (Riki Donley Saxton Collins III)
before the Great Selenite Find in 1962

Bibliography

Eckel, E.B., 1950, Minerals of Colorado—A Hundred Year Record, USGS Bulletin 1114, 399 p.

Pearl, Richard M., 1958, Colorado Gem Trails and Mineral Guide, 3d edition, Sage Books, Denver, Colorado, 176p.

AUTHOR

The author, Donley Collins, is a lifetime resident of Colorado who began collecting minerals in 1956 at the age of 8. He was born in Colorado Springs and spent most of his mineral collecting in the surrounding area. While a member of the Colorado Springs Mineralogical Society, he was the youngest to be a field trip leader. In 1979, he found the first diamond, graphite (type 2) eclogite in North and South America and has since contributed to and is cited in many articles. Today at the age of 76, he continues to search mineral shows for improperly labeled and new mineral discoveries as well as cleaning the many minerals collected over the last 45 years. He earned a Bachelor of Science Degree (BS) at Colorado State University and a master's degree in Geology from the Colorado School of Mines. For 20 years he was a research geologist for the United States Geological Survey (USGS)

specializing in stratigraphy, geomorphology, and sedimentary and igneous petrology. Based on his knowledge and interest in Colorado mineralogy, he was among the first requested by E. B. Eckel to update the popular book "Minerals of Colorado A 100-Year Record". USGS Bulletin 1114 published in 1961.

EDITOR

The editor, Bob King, has been digging for crystals for 63 years. He just "loves" to dig and at the age of 87 he still does. His minerals have been exhibited competitively in many mineral shows across the USA and Australia. While he and his family were living in Australia (1972-1974) he was instrumental in the organizing of the Echuca Gem Club in which he is a Lifetime Member. Obtaining "Miner's Rights" licenses in two states in Australia gave him permission to collect on "Crown Land" in those states, which he did. Here in the USA, he has dug in 18 different states, mostly in Colorado and Wyoming. Bob was the president of the Cheyenne Mineral & Gem Society for ten years.

In 2008 Bob was the first recipient to receive the "Rockhound of the Year" award in the State of

Wyoming. This award is "In recognition of his outstanding contributions to the hobby of rockhounding." He currently lives in Cheyenne, Wyoming, but as a Lifetime Member of the Colorado Springs Mineral Society he keeps in touch with that club.

Photo Acknowledgements

All photos are from the collection by Donley S. Collins except for the following:

Fisher, George W.	Photo 30
King, Robert W.	Photo 18,34,40,48a,53
Kowalski, Andy	Photo 10,12,41a,b,c, 51,55, 67